MW00963250

BASIC / NOT BORING

NUMBERS & COUNTING

Grades K–1

Inventive Exercises to Sharpen Skills and Raise Achievement

Series Concept & Development
by Imogene Forte & Marjorie Frank
Exercises by Sharon Gell

Incentive Publications, Inc.
Nashville, Tennessee

About the cover:

Bound resist, or tie dye, is the most ancient known method of fabric surface design. The brilliance of the basic tie dye design on this cover reflects the possibilities that emerge from the mastery of basic skills.

Illustrated by Kathleen Bullock
Cover art by Mary Patricia Deprez, dba Tye Dye Mary®
Cover design by Marta Drayton, Joe Shibley, and W. Paul Nance
Edited by Anna Quinn

ISBN 0-86530-386-X

PRINTED IN THE UNITED STATES OF AMERICA

TABLE OF CONTENTS

Appendix

CELEBRATE BASIC MATH SKILLS

Basic does not mean boring! There is certainly nothing dull about . . .

. . . counting jump-roping bunnies, jogging dinosaurs, and bouncing frogs
. . . walking a number line with a bunch of ladybugs
. . . catching numbers (and catfish) on a fishing line
. . . turning pogo stick bounces into numerals
. . . figuring out how many campers are stuffed into their tents
. . . using numbers to help swimmers get ready for a race
. . . comparing numbers of skiers crashing on the ski slope
. . . helping goofy golfing gophers keep track of their scores
. . . putting numbers in order to locate a monkey who's far from the jungle

The idea of celebrating the basics is just what it sounds like—enjoying and improving the skills of counting and using numbers. Each page of this book invites young learners to try a high-interest, visually appealing exercise that will sharpen one specific math skill. This is not just any ordinary fill-in-the-blanks way to learn. These exercises are fun and surprising, and they make good use of thinking skills. Students will do the useful work of practicing math skills while they enjoy exciting sports and recreation adventures with animals.

The book can be used in many ways:

- to review or practice a math skill with one student
- to sharpen the skill with a small or large group
- to start off a lesson on a particular skill
- to assess how well a student has mastered a skill

Each page has directions that are written simply. It is intended that an adult be available to help students read the information on the page, if help is needed. In most cases, the pages will best be used as a follow-up to a lesson or concept that has been taught. The pages are excellent tools for immediate reinforcement of a concept.

As your students take on the challenges of these adventures with math, they will grow! And as you watch them check off the basic math skills they've acquired or strengthened, you can celebrate with them.

The Skills Test

Use the skills test beginning on page 58 as a pretest and/or a post-test. This will help you check the students' mastery of counting and numbers. You'll find out what they've learned or what they need to practice.

SKILLS CHECKLIST
NUMBERS & COUNTING, GRADES K-1

✔	SKILL	PAGE(S)
	Count to numbers less than 100	10–12, 14, 15, 31
	Count to 100	13, 16
	Skip count by 2	14
	Skip count by 5	15
	Skip count by 10	16
	Read and use ordinals	17
	Match numbers to sets and models	18, 19
	Read and write numbers on number lines	20, 21
	Match numerals to word names	22, 23
	Writing and ordering numbers	24
	Read and write whole numbers to 5 digits	24, 38
	Identify place value through hundreds	25–29
	Read and write numbers in expanded form	25, 28, 29
	Compare amounts using comparison words	30, 31, 34, 43, 54, 55
	Compare whole numbers using < and >	31–35
	Order whole numbers	36
	Round numbers to the nearest ten	37
	Explore big numbers	38
	Estimate numbers of objects	39
	Identify values of coins	40–42
	Count money	41, 42
	Write and compare amounts of money	43
	Match fractions to pictures and models	44–46
	Read, write, and illustrate simple fractions	44–46
	Compare fractions	46
	Tell time to the half hour	47–49
	Find numbers and information on a calendar	50
	Measure length	51
	Recognize units for measuring length, weight, and capacity	51–55
	Compare measurements	54–55

NUMBERS & COUNTING

Grades K–1

Skills Exercises

More & More Marbles

Do you think little dinosaurs play marbles in their caves?

Explore the cave.

Find all the marbles, and color them different colors.

Count the marbles. How many did you find? ______________

Name ______________________________

Fun Run

Take a run with these joggers.

Watch out for the little animals on the path!

Color all the animals, and then count them.

1. How many  ? ☐

2. How many  ? ☐

3. How many  ? ☐

4. How many  ? ☐

5. How many  ? 

6. How many  ? ☐

Name ______________________________

Splash Down

It's Turtle Fun Day at Water Slide Park.

Color the ladder different colors.

Walk your fingers up the ladder to count the steps.

How many are there? ☐ Then slide down with the turtles.

Go up again. How many steps did you count this time? ☐

Color the turtles.

Name ______________________________

Trampoline Trouble

Fran and Felix want to jump **100** times on the trampoline.

Write in the missing numbers so they can jump on each number.

Use your fingers to jump to each number.

Say the numbers as you jump.

If you make a mistake, climb on and try again.

1	2	3		5	6	7			10
11		13	14	15		17	18	19	
21	22		24	25		27		29	30
31		33	34		36		38		40
41			44	45		47		49	50
	52	53	54		56	57	58		
61		63		65	66		68	69	70
	72		74	75				79	
81	82	83		85	86	87		89	90
91	92		94			97	98		100

Color Fran and Felix.

Name ______________________________

Double Doubles

The Bunny Hop twins do everything by twos.

What are they doing today?

Count by twos, and draw a line from one number to the next.

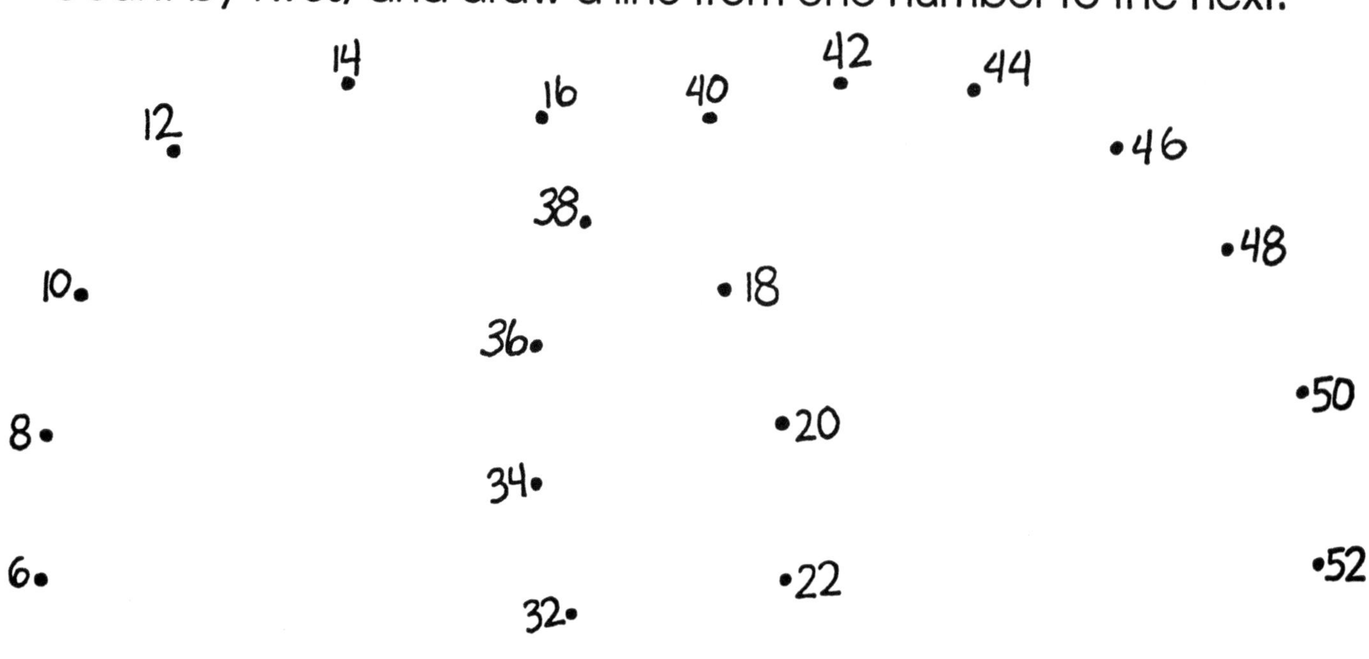

Color the twins. How high did you count? ☐

Name ____________________

High Fives

What a game! Each player bounces the ball 5 times.

Count by fives.

Write the missing numbers on the players' shirts.

What is the highest number you counted? ☐

Which 2 players are making a "high five"? ☐ and ☐

Color the picture.

Name ____________________

100-Yard Dash

Kerry Kangaroo always hops by tens.

Help her hop 10 times to get to 100.

Count by tens and write in the missing numbers.

Color Kerry's path.

Name ____________________

The Winner Is . . .

The race is almost over.

The first 5 runners are crossing the finish line.

Draw a line from each runner to the ribbon he will get.

Color the winning turtle green.

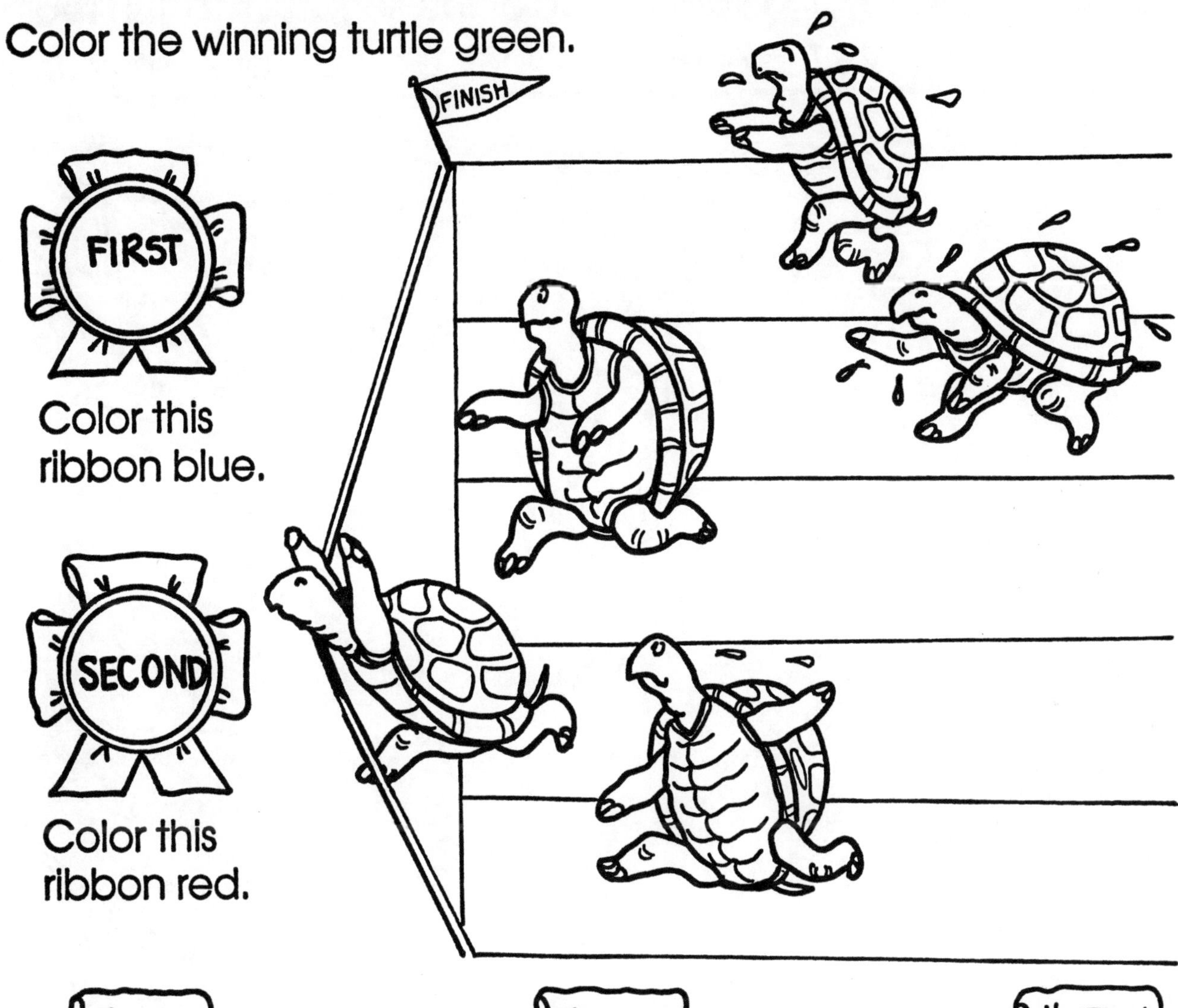

Color this ribbon blue.

Color this ribbon red.

Color this ribbon yellow.

Color this ribbon green.

Color this ribbon purple.

Name ______________________________

Under the Sea

Help Diver Dan count the fish in 4 schools!

A school of fish is a group of the same kind of fish that swim together.

Color the school of **11** fish **yellow.** Color the school of **21** fish **red.**
Color the school of **19** fish **purple.** Color the school of **7** fish **orange.**

Name ___________________________

A Messy Room

Abby's sports things are all over her room.
Some things are missing.
Read the total number for each group.
Draw the missing things.

4 footballs

2 baseball caps

11 hockey pucks

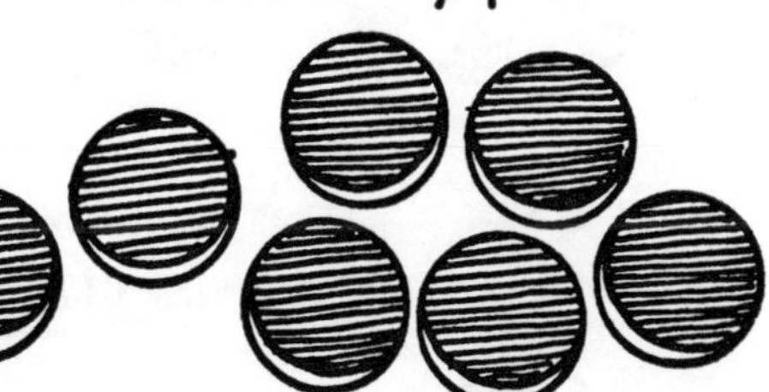

6 hockey sticks

3 baseball bats

7 golf balls

Color Abby.

Name ___________________________________

Long Jumpers

Three very good jumpers are having a contest.

Who can jump the farthest?

Look at the number line to see how far each one has jumped.

1. jumped ☐ feet.
2. jumped ☐ feet.
3. jumped ☐ feet.

Draw yourself on the number line jumping 25 feet.

Did you jump farther than the ? **yes** **no**

Did jump farther than the ? **yes** **no**

Did jump farther than the ? **yes** **no**

Color all the jumpers.

Name ____________________________________

Ladybug Marathon

The ladybugs make a nice number line.

Use your fingers to walk the number line.

Answer each ladybug's question when you land on her.

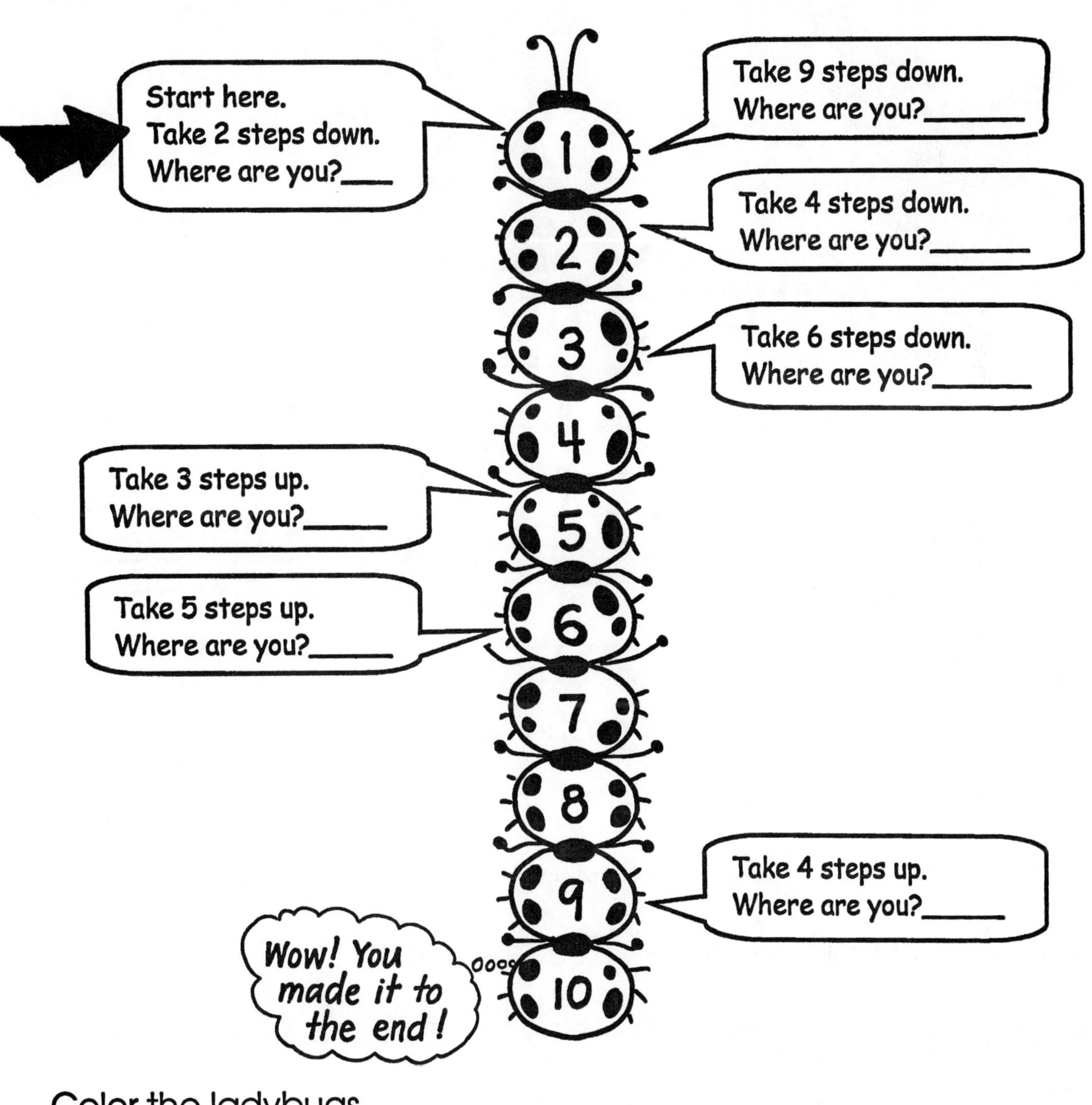

Color the ladybugs.

Name ______________________________

Swim for It

At the Animal Olympics, 6 swimmers are lined up for the big race.
Help them get ready to swim!
Follow the directions.

1. Draw goggles on the swimmer in lane three.
2. Color the swimmer in lane five yellow.
3. Draw teeth  on the swimmer in lane one.
4. Make the swimmer in lane six come out of his shell.
5. Draw a hat  on the swimmer in lane two.
6. Color the swimmer in lane four brown.

Name ______________________________

Too Many Players

A soccer team has eleven players.

Does this team have the right number?______

Read the numbers on the shirts.

Write the numbers beside the matching words.

1. ☐ fifteen
2. ☐ twelve
3. ☐ seventy-one
4. ☐ thirty-three
5. ☐ thirteen
6. ☐ fifty-one
7. ☐ ninety
8. ☐ sixty-two
9. ☐ seventeen
10. ☐ eight
11. ☐ fifty
12. ☐ twenty-six

Name ______________________________

Fishing for Sixes

Cousin Cary Cat is fishing for catfish.
He has 6 fat catfish on his line.
How many pounds does each fish weigh?
Write the correct number on each fish.

Color the heaviest fish orange.
Color the other fish any colors.

Name ______________________________

A Long, Hot Run

These runners are very thirsty.
They have run a long way on a hot day.
They need to drink a lot of water.

Bonnie ran for 147 minutes.	147 = 100 + 40 + 7
Barney ran for 199 minutes.	199 = 100 + 90 + 9
Ben ran for 84 minutes.	84 = 80 + 4

Look at the times for the other runners. Draw a line to the glass of water that shows what the number means.

1. 89 minutes
2. 96 minutes
3. 133 minutes

4. 75 minutes
5. 69 minutes
6. 98 minutes

Color the runners.

Name ______________________________

A Pogo Puzzle

Polly and Pagoo are on a pogo stick team.
They bounce many times.
The words tell how many times they bounce.
Write the numbers to match the words.
Then color Polly and Pagoo.

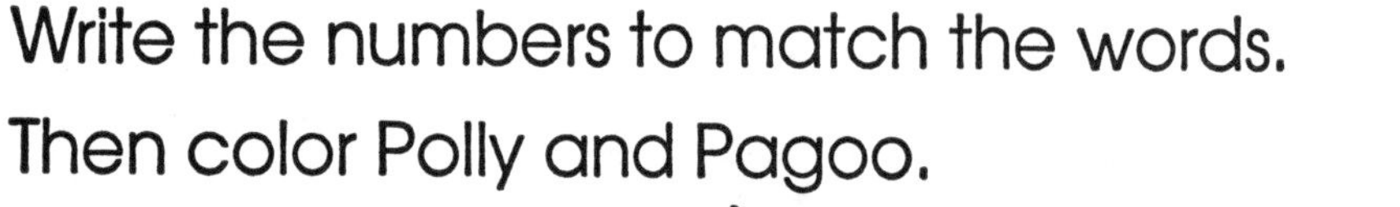

Across

A. 2 tens, 3 ones
C. 1 hundred, 0 tens, 0 ones
D. 8 tens, 7 ones
E. 6 tens, 5 ones
F. 9 tens, 2 ones
G. 4 tens, 1 one
H. 6 tens, 0 ones

Down

A. 2 tens, 0 ones
B. 3 tens, 0 ones
C. 1 ten, 7 ones
D. 8 tens, 5 ones
E. 6 tens, 6 ones
F. 9 tens, 1 one
G. 4 tens, 0 ones

Name ____________________________

A Hidden Sport

Sara is happy about a game she has won.
But what is her sport?
A clue is hiding in the puzzle.
Follow the code to color the spaces.
Find out what is hiding.

COLOR CODE

Look at the tens place in each number.

2 or **6** in the tens place = **black**
4 or **7** in the tens place = **green**
5 or **8** in the tens place = **red**
1 or **3** in the tens place = **yellow**

Name ___________________________

Full Tents

This is the best camping trip of the year.

The tents are full of campers.

The number on each tent tells how many are inside.

Write the tent number that means the same as each of these.

1. 20 + 3 = ☐
2. 80 + 8 = ☐
3. 40 + 7 = ☐
4. 20 + 6 = ☐
5. 10 + 6 = ☐
6. 30 + 0 = ☐

7. Which number has the most tens? ☐
8. Which number has the fewest tens? ☐
9. Which number has the most ones? ☐
10. Which number has the fewest ones? ☐

Color the picture.

Name ______________________________

A Wild Tennis Game

This tennis game has gotten very wild.
There are tennis balls everywhere!
Look at the numbers on the balls.
Follow the directions to color the balls.

11

90

45

Find the ball that means the same as each of these, and color it.

30 + 2 . . . . . red	10 + 1 . . . . . blue	40 + 5 . . . . pink
10 + 4 . . . . . blue	80 + 4 . . . . . green	100 + 20 + 3 . . . red
90 + 0 . . . . . purple	50 + 0 . . . . . green	20 + 9 . . . . orange
60 + 6 . . . . . brown	70 + 6 . . . . . yellow	100 + 10 + 1 . . . purple

Name ____

Hula Hoop Dancers

The hula hoop dancers are dancing on the sand.
They dance and spin 1 or more hoops.
But where are their hoops?

Follow the directions to draw the hoops.

1. Draw a red hula hoop on the shortest dancer.
2. Draw a blue hula hoop on the dancer with the longest skirt.
3. Draw an orange hula hoop on the dancer with the most coconuts.
4. Draw a green hula hoop on the dancer with the fewest necklaces.
5. Draw a purple hula hoop on the dancer with the most necklaces.
6. Draw a yellow hula hoop on the dancer with the fewest coconuts.

Name ____________________

Headed for Trouble

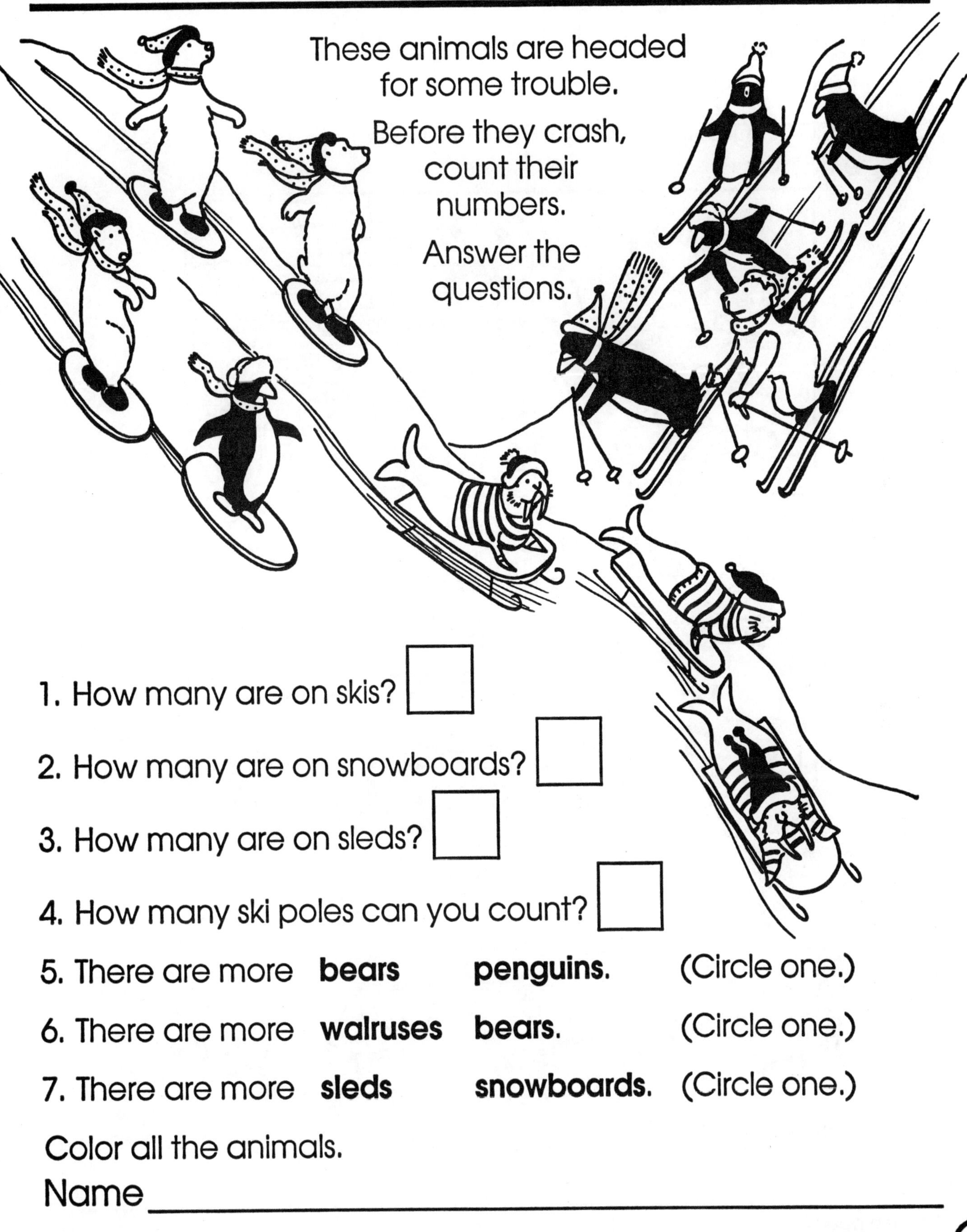

1. How many are on skis? ☐
2. How many are on snowboards? ☐
3. How many are on sleds? ☐
4. How many ski poles can you count? ☐
5. There are more **bears** **penguins.** (Circle one.)
6. There are more **walruses** **bears.** (Circle one.)
7. There are more **sleds** **snowboards.** (Circle one.)

Color all the animals.

Name ______________________________

On the Shelves

Mr. Stork has to count all the equipment in his store every year. It's a big job. He needs your help.

Count each group, and write the number on the next page (page 33).

Color all the things on the store shelves.

Name ______________________________

Use with page 33.

On the Shelves, continued . . .

Write the number of each kind of thing you counted.

< means **less than**

> means **greater than**

= means **equal to**

Put < or > in each box.

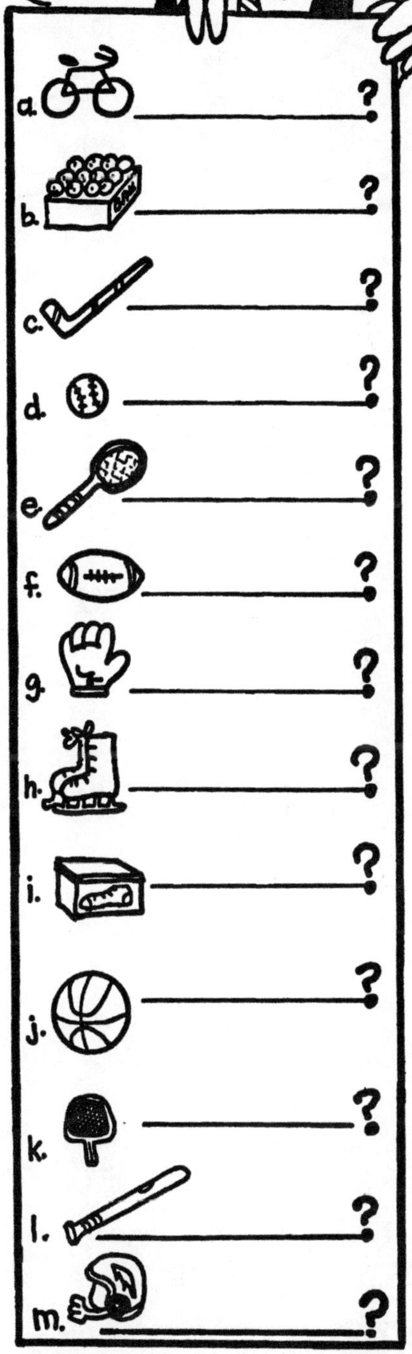

1. ☐
2. ☐
3. ☐
4. ☐
5. ☐
6. ☐
7. ☐
8. ☐
9. ☐
10. ☐
11. ☐

a. ________ ?

b. ________ ?

c. ________ ?

d. ________ ?

e. ________ ?

f. ________ ?

g. ________ ?

h. ________ ?

i. ________ ?

j. ________ ?

k. ________ ?

l. ________ ?

m. ________ ?

Name ____________________

Use with page 32.

Goofy Golfing Gophers

Golfers count the number of times they hit the ball for each hole. The golfer with the lowest number of hits is the winner.

Count the number of hits each golfer takes to get the ball into this hole.

Write the numbers on the scorecard.

SCORECARD	
Gus........	
Gordy......	
Gail........	
Gabby......	
Gipper.....	

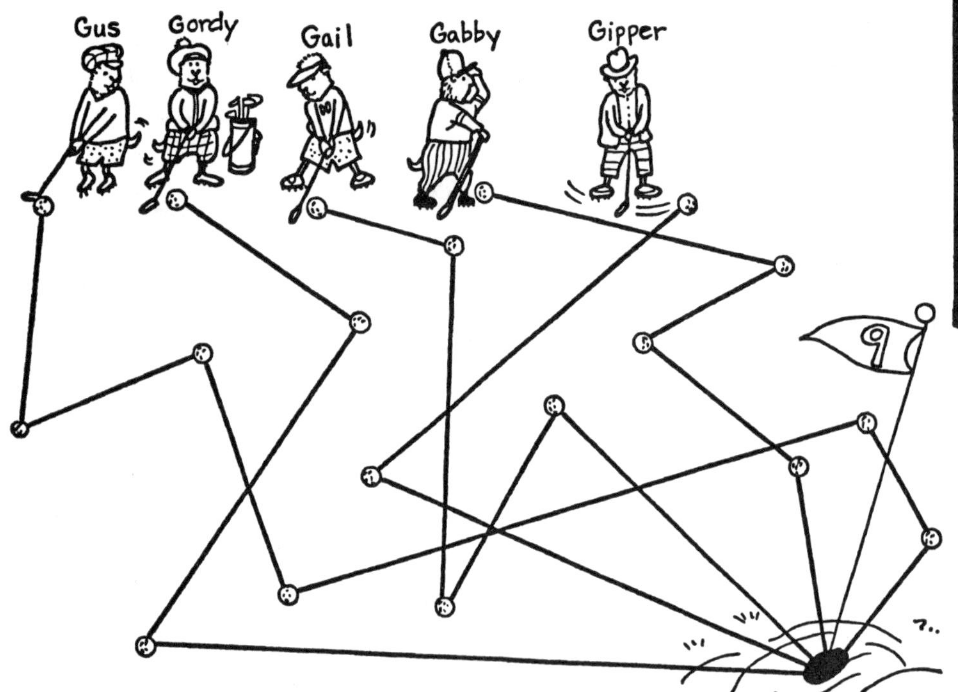

1. Who took more hits than Gabby? ____________________

2. Who took less hits than Gordy? ____________________

3. Who took the most hits on this hole? ____________________

4. Who is the winner at this hole? ____________________

Color the golfers.

Name ______________________________

Use with page 35.

Goofy Golfing Gophers, continued . . .

At the end of the day, the goofy gophers added up all their scores. Look at the scorecard to compare their scores.

< means **less than**

> means **greater than**

SCORECARD	
Gus	50
Gabby	68
Gail	104
Gordy	42
Gipper	66
Greta	114

Then write < or > in each box.

1.	Gordy		Gus
2.	Gail		Gordy
3.	Gabby		Gus
4.	Gordy		Greta
5.	Gail		Gipper
6.	Gus		Gail
7.	Gipper		Gabby
8.	Greta		Gabby
9.	Gipper		Gus
10.	Gordy		Gabby

Color the golfers.

Name ______________________________

Use with page 34.

Monkey Business

This monkey is a long way from his home in the jungle.
Connect the dots to find out where he is and what he is doing.
Connect the dots by putting the numbers in order from 1 to 38.

Color the picture.

Name ______________________________

Round-up Rodeo

Get out your lasso!
Join the round-up rodeo.
Help Cowboy Carl round the numbers.

Color all the 10 numbers blue.
Pick 10 numbers that are not colored.
Write each number on a line below.
Round each number to the nearest 10.

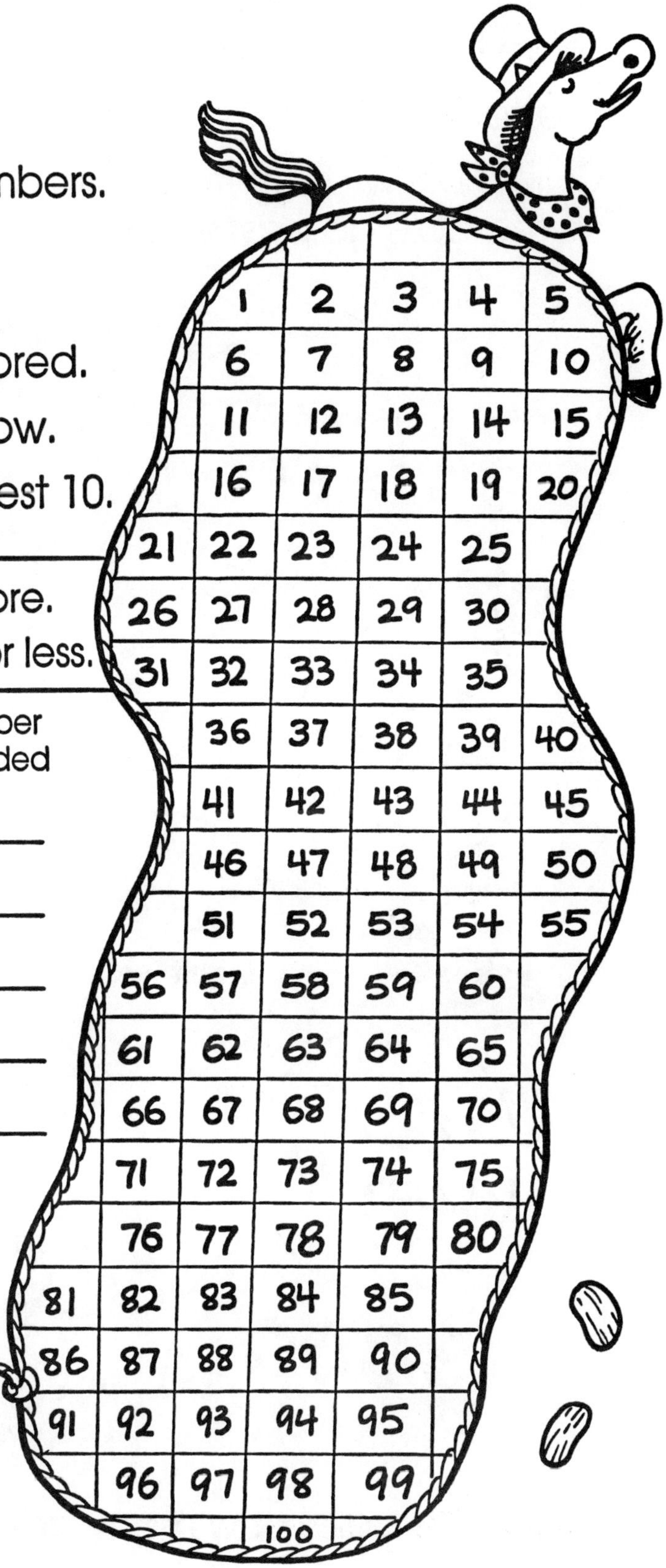

Round **UP** numbers that end in **5** or more.
Round **DOWN** numbers that end in **4** or less.

	Number	Number Rounded		Number	Number Rounded
1.	______	______	6.	______	______
2.	______	______	7.	______	______
3.	______	______	8.	______	______
4.	______	______	9.	______	______
5.	______	______	10.	______	______

1	2	3	4	5
6	7	8	9	10
11	12	13	14	15
16	17	18	19	20
21	22	23	24	25
26	27	28	29	30
31	32	33	34	35
36	37	38	39	40
41	42	43	44	45
46	47	48	49	50
51	52	53	54	55
56	57	58	59	60
61	62	63	64	65
66	67	68	69	70
71	72	73	74	75
76	77	78	79	80
81	82	83	84	85
86	87	88	89	90
91	92	93	94	95
96	97	98	99	
		100		

Color the cowboy and his horse.

Name ______________________________

Table Tennis Terror

There has been an accident at the table tennis ball factory.

The table tennis balls are bursting out the door!

Quick! Draw a line from each ball to the matching number as the balls go past you.

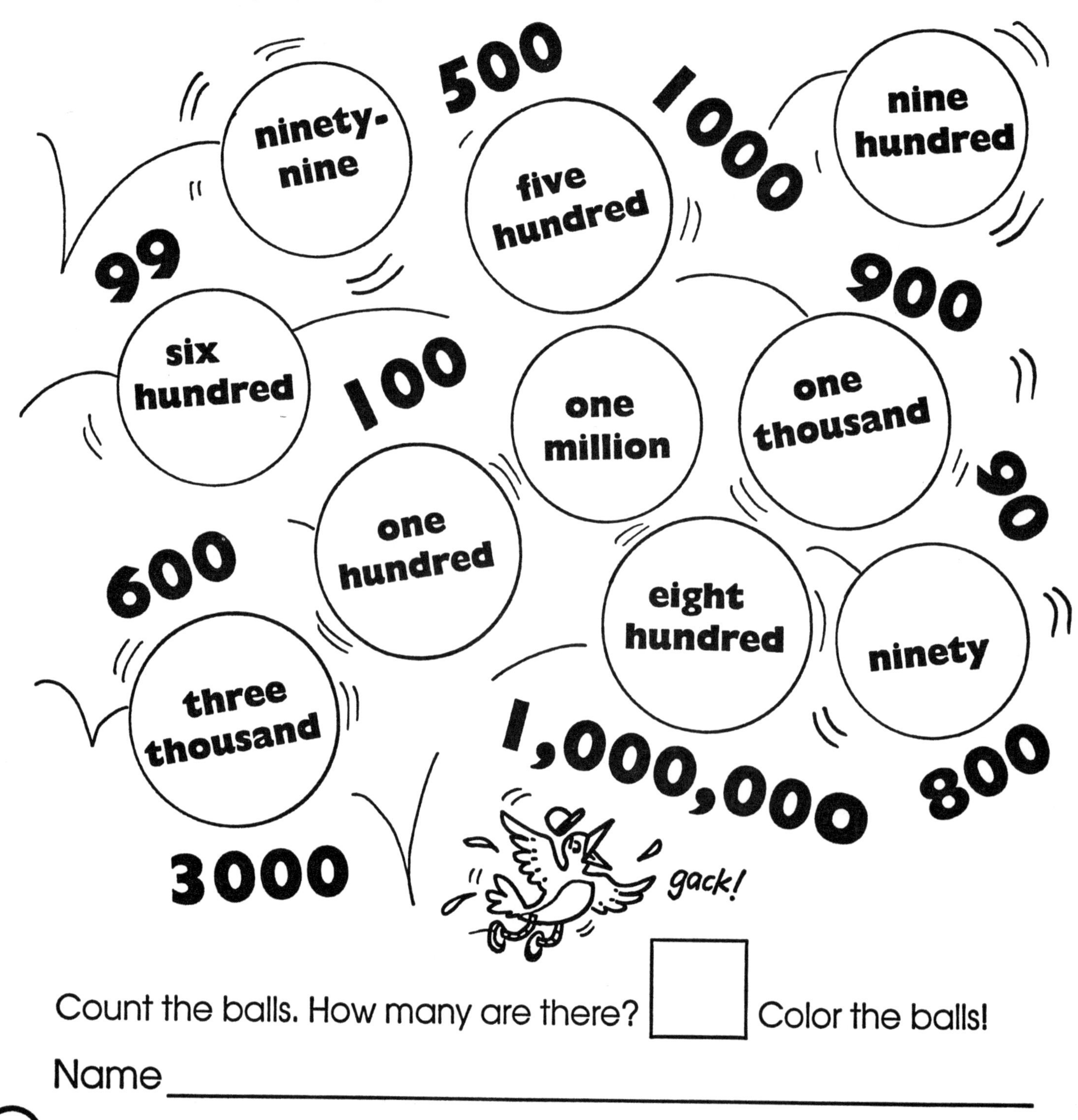

Count the balls. How many are there? ☐ Color the balls!

Name ______________________________

Hide-and-Seek

The Mousekin kids are playing hide-and-seek in their room. Take a quick look at the picture. Then close your eyes. Estimate how many hiding places the mice have found.

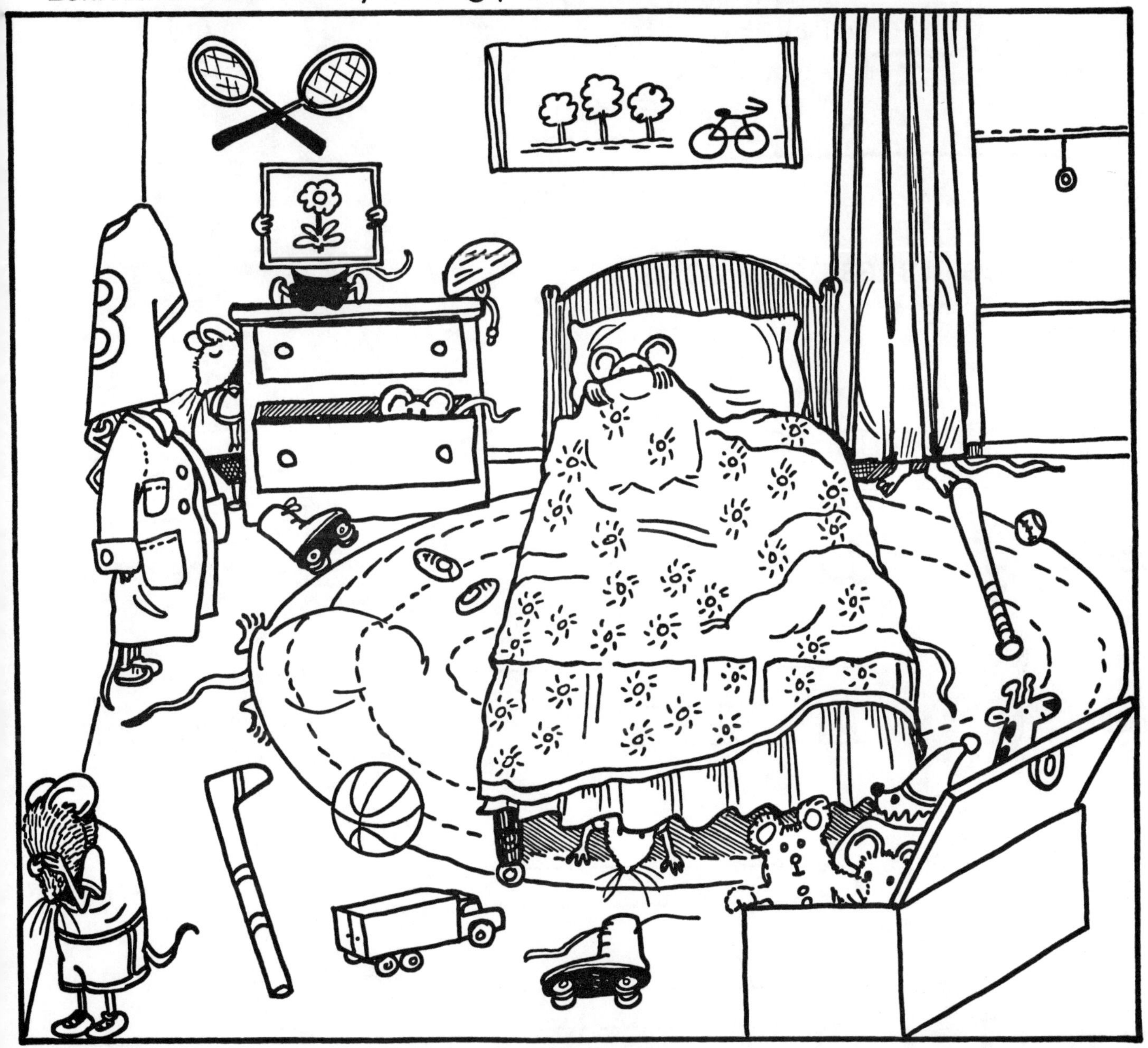

How many did you estimate? ☐ Now count the hidden mice.

How many did you find? ☐ Color the picture.

Name ____________________

Find a Penny

Lucky Penny likes to search the beach for lost coins.
She puts the coins in her piggy banks.

"Find a penny
Pick it up
All the day
You'll have Good Luck!"

Write the names of the coins on the lines.

Write the amount for each coin on the piggy bank.

penny p ____ ¢ one cent

nickel n ____ ¢ five cents

dime d ____ ¢ ten cents

quarter q ____ ¢ twenty-five cents

Color Lucky Penny.

Name ____________________

Use with page 41.

Find a Penny, continued . . .

Count the money.

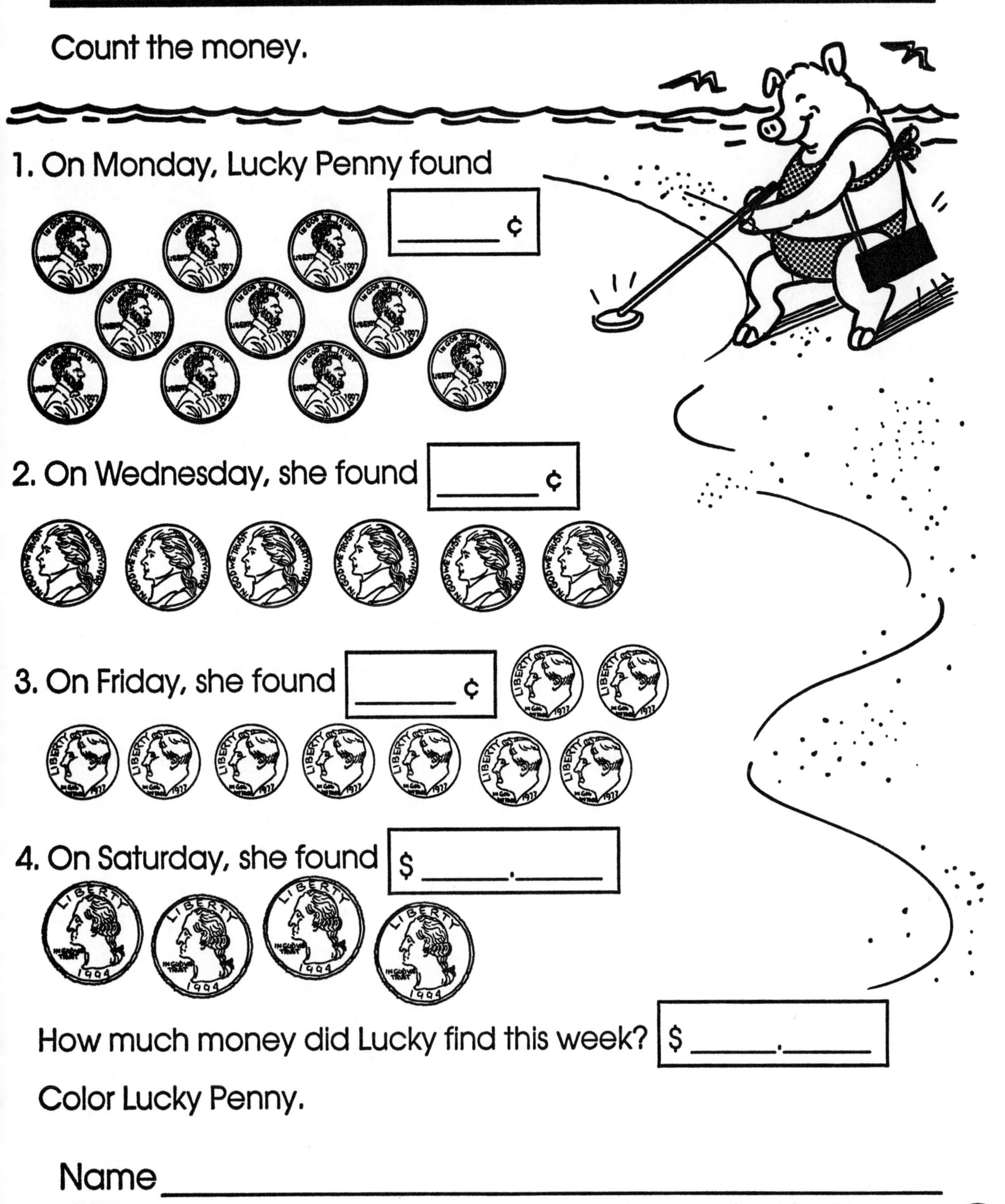

1. On Monday, Lucky Penny found ______ ¢

2. On Wednesday, she found ______ ¢

3. On Friday, she found ______ ¢

4. On Saturday, she found $ _____._____

How much money did Lucky find this week? $ _____._____

Color Lucky Penny.

Name ________________________________

Use with page 40.

Under the Bleachers

The friends found money under the bleachers at the ball game.
Circle the correct names of the coins they found.
Count the money, and write the amount on the line.

1. 

Max found 2
pennies
dimes
quarters
_______¢

2. 

Tracy found 6
dimes
pennies
nickels
_______¢

3. 

Ollie found 1
quarter
dime
penny
_______¢

4. 

Ben found 4
nickels
dimes
quarters
$_______.___

5. 

Annie found 1
dime
quarter
penny
and 1
nickel
penny
quarter
_______¢

6. 

Ellen found 2
dimes
nickels
pennies
_______¢

Color the friends.
Who found the most money? ____________________

Name ______________________________

Snacks at the Ball Game

Simon Skunk sells snacks at softball games.
Look at all the yummy snacks in his tray.

Circle **more** or **less.**

1. costs **more** **less** than
2. costs **more** **less** than
3. costs **more** **less** than
4. costs **more** **less** than
5. costs **more** **less** than
6. costs **more** **less** than

Color Simon and his snacks.

Name ______________________________

A Very Hungry Hockey Player

Hannah is very hungry after the hockey game.
She goes to the pizza shop and eats lots of pizza.
Help the pizza shop put toppings on the pizza.

1. Draw pepperoni on one half.

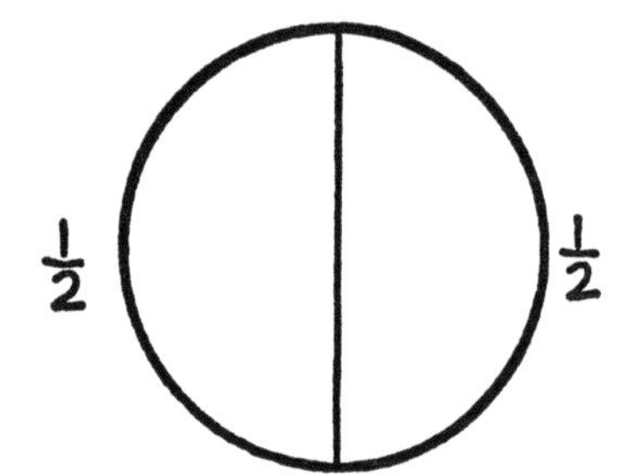

2. Draw pineapple on one quarter.

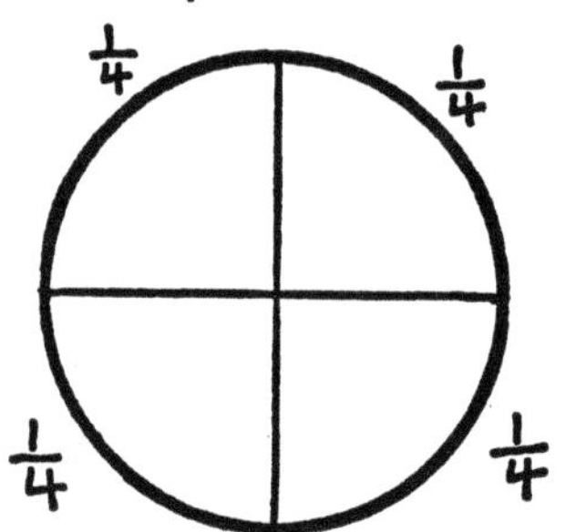

3. Draw olives on one eighth.

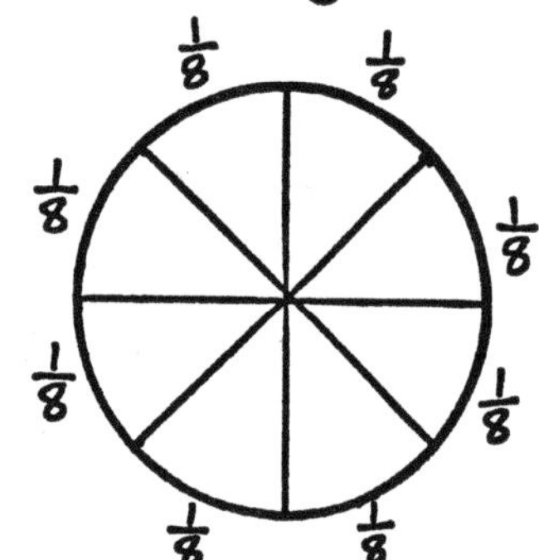

4. Draw your favorite toppings on one sixth.

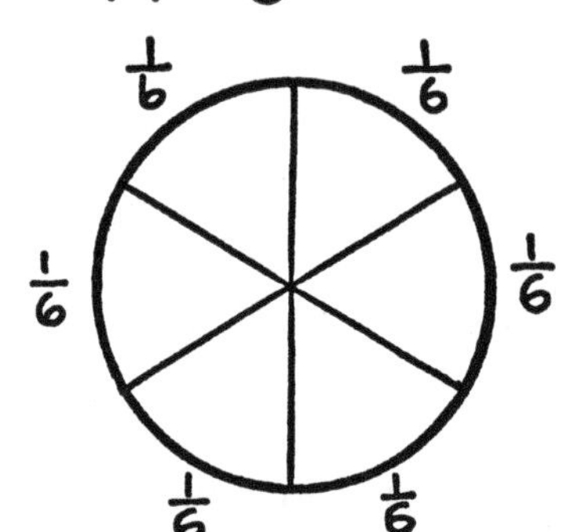

Name ______________________________

The 3-Legged Race

Help the mice find their trails for the 3-legged race.

Millie and Mary follow a trail of halves.

Use a blue crayon to connect all the ways to show $\frac{1}{2}$.

Mike and Mac follow a trail of fourths.

Use a red crayon to connect all the ways to show $\frac{1}{4}$.

THREE-LEGGED RACE

one half

one quarter

one fourth

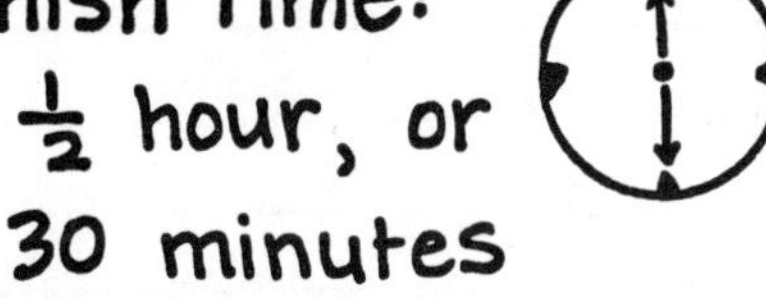

Finish Time:
$\frac{1}{4}$ hour, or
15 minutes

Color the mice.

Name ____

First Prize for Yummy Pies

Today is the pie-eating contest.

The pies are yummy, and the pie eaters are very full.

Look at the sizes of the pie pieces that are left.

1. Draw a box around the fraction that shows the biggest piece.
2. Draw a circle around the fraction that shows the smallest piece.
3. What is your favorite kind of pie? ______________________
4. Which size piece would you like to eat? ______________________

Color the picture.

Name ______________________

They're Off!

The horses are ready!

The clock shows what time the race will start.

Write the missing numbers on the clock.

Then complete the sentences.

12

9 3

6

On your mark, get set...

HEY! He didn't say "Go" yet!

I'm off!

1. The minute hand is on the ☐.

2. The hour hand is on the ☐.

3. The time is ☐ : ☐.

4. In $\frac{1}{2}$ hour it will be ☐ : ☐.

Color the picture.

Name ______________________________

Snow Time

Snow days are fun from morning until night.
Kids can think of fun things to do all day long.
Write the time that is shown on each clock.

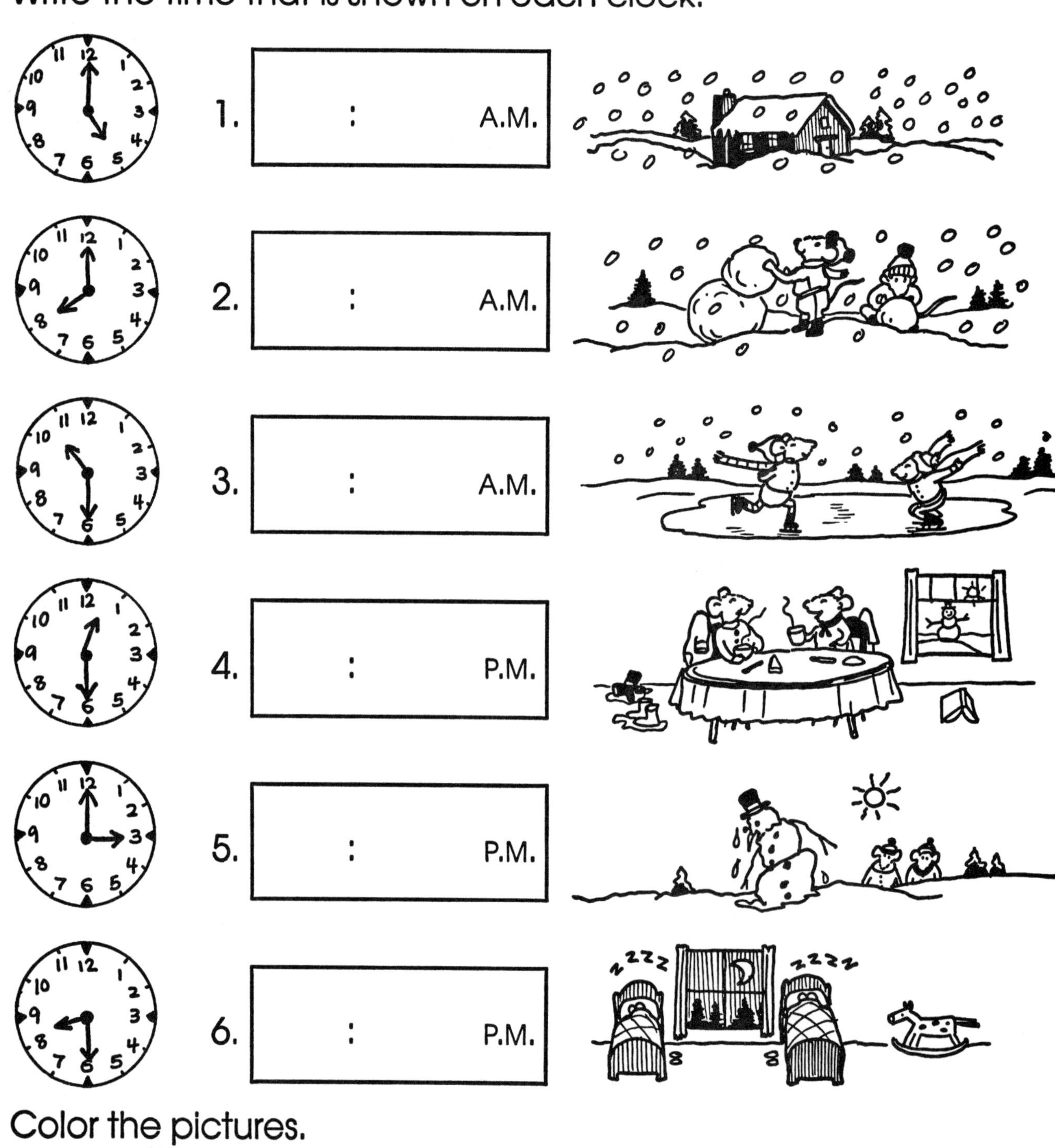

Color the pictures.

Name ____

Show Time

The shows are starting! Don't be late!

Read about each show, and follow the directions.

1. The marching band starts a show at 3:00. It lasts 30 minutes.
   Draw the hands on the clock to show the time it will end.

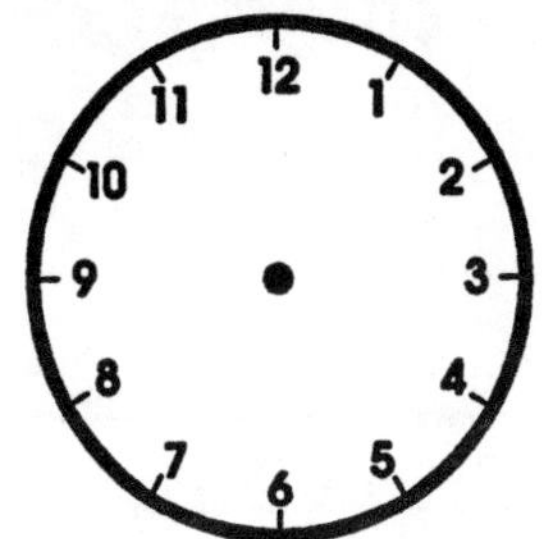

2. The rodeo starts at 4:00. It lasts for 1 hour.
   Draw the hands on the clock to show the time it will end.

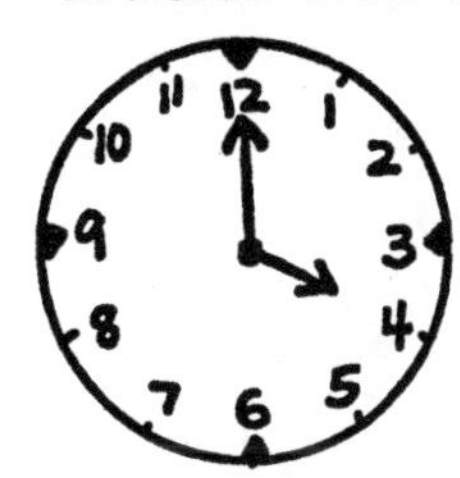

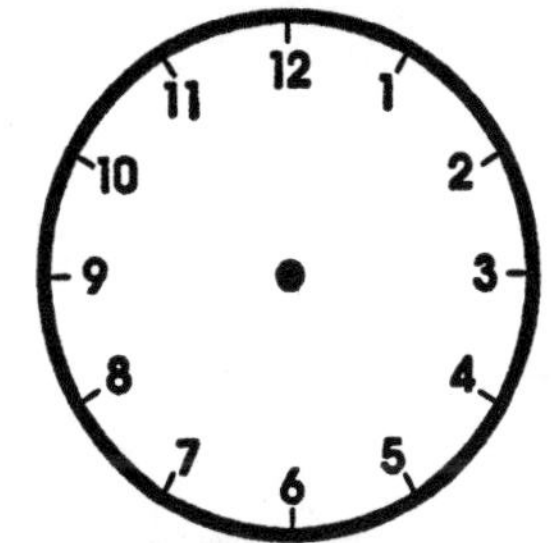

3. The ice show starts at 5:30. It lasts for 1 hour.
   Draw the hands on the clock to show the time it will end.

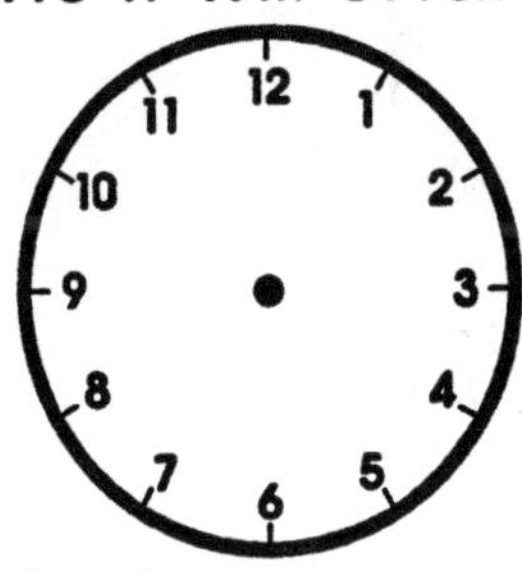

4. The water show starts at 7:00. It lasts for 30 minutes.
   Draw the hands on the clock to show the time it will end.

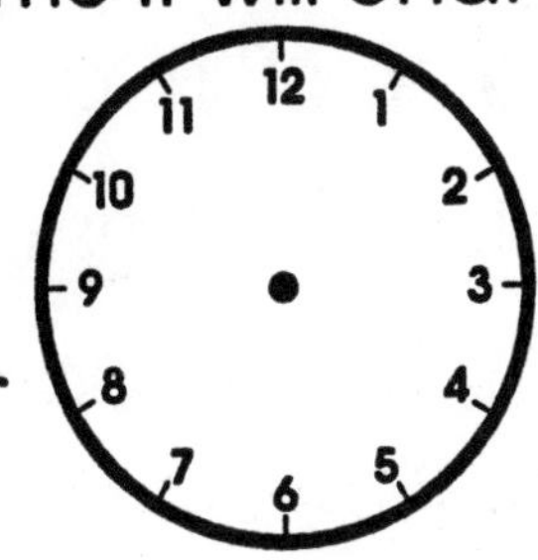

Color the pictures.

Name ______________________________

Fantastic Gymnastics

Ari loves gymnastics.
He goes to the gym every Tuesday to practice.
He practices some other days, too.

Look at Ari's calendar.
Fill in the missing numbers.

FEBRUARY						
Sun.	Mon.	Tues	Wed.	Thurs.	Fri.	Sat.
1	2	3 gym		5	gym	7
8	9	10 gym	11		13	
15		gym	18	gym	20	gym
22		24 gym		meet	27	

1. Write the dates of all the Tuesdays. ____, ____, ____, ____
2. How many Tuesdays are there? ____________
3. How many days are there from one Tuesday to the next? (Count each day of the week only once.) ______
4. Is the gymnastics meet on a Wednesday? **yes no** (Circle one.)
5. Valentine's Day is February 14th. Draw a heart on that day.

Color Ari.

Name ______________________________

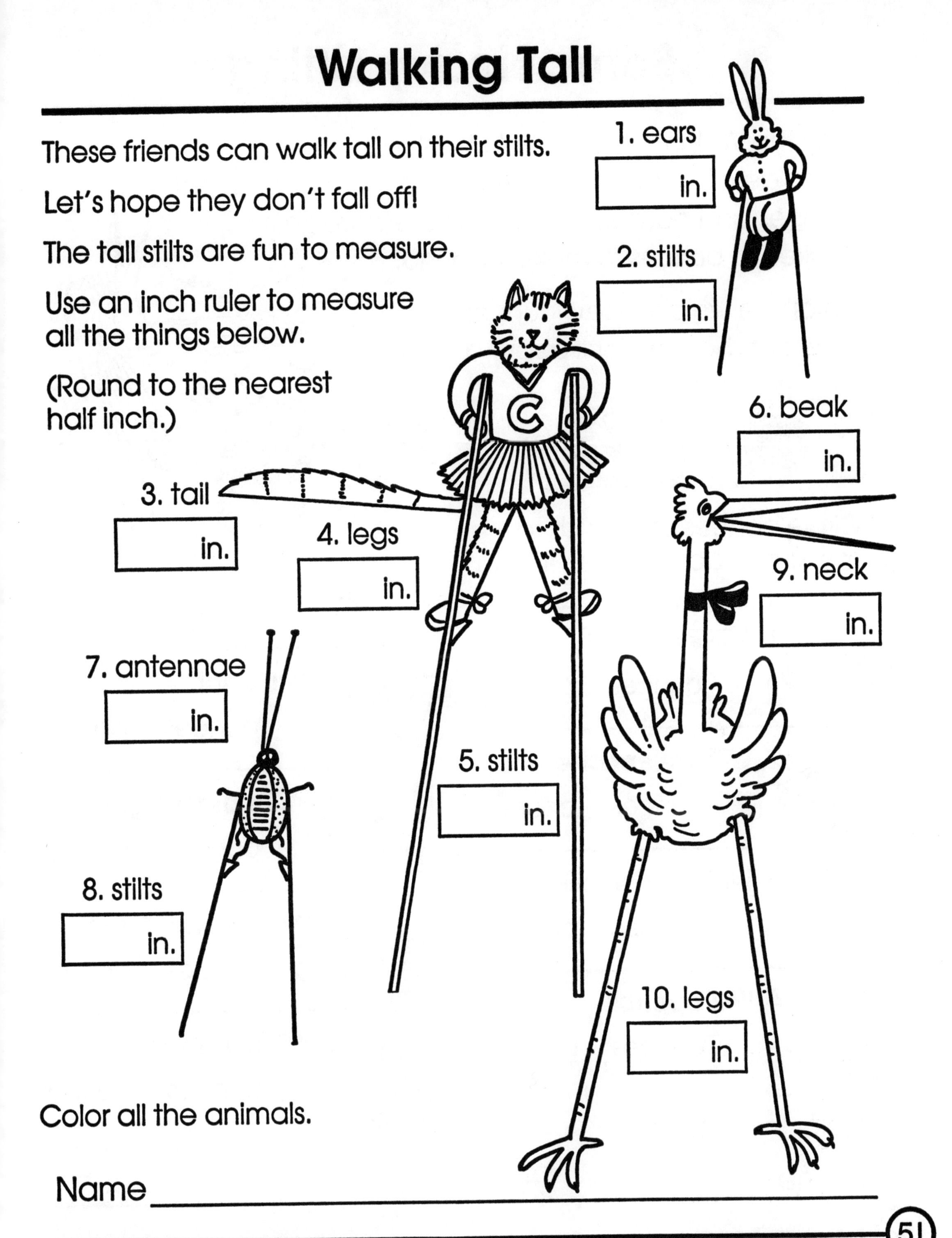
Walking Tall
These friends can walk tall on their stilts.
Let's hope they don't fall off!
The tall stilts are fun to measure.
Use an inch ruler to measure all the things below.
(Round to the nearest half inch.)
1. ears
in.
2. stilts
in.
6. beak
in.
3. tail
in.
4. legs
in.
9. neck
in.
7. antennae
in.
5. stilts
in.
8. stilts
in.
10. legs
in.
Color all the animals.
Name

Some Heavy Lifting

Willy the Weight Lifter is a little lopsided.
One side of his bar has **ounces,** and the other has **pounds.**
Ounces and pounds are very different weights.
It takes 16 ounces to make 1 pound.

An apple weighs a few ounces.
A pair of shoes weighs about a pound.

Circle ounces or pounds to measure each of these things.

1. **Will's cookies**
   ounces pounds

5. **Will's hat**
   ounces pounds

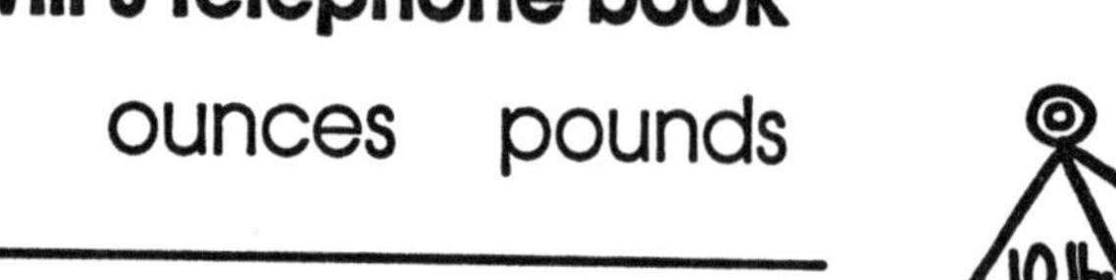

2. **Will's telephone book**
   ounces pounds

6. **Will's big weight**
   ounces pounds

3. **Will's jug of milk**
   ounces pounds

7. **Will's friend, Oscar**
   ounces pounds

4. **Will's pencil**
   ounces pounds

8. **Will's grapes**
   ounces pounds

Color the pictures.

Name ______________________________

A Great Sand Castle

Sandy Squirrel has built a sand castle for a contest at the beach. She deserves a blue ribbon for the most beautiful sand castle. What did she use to make the castle?
She used a cup and a pail.

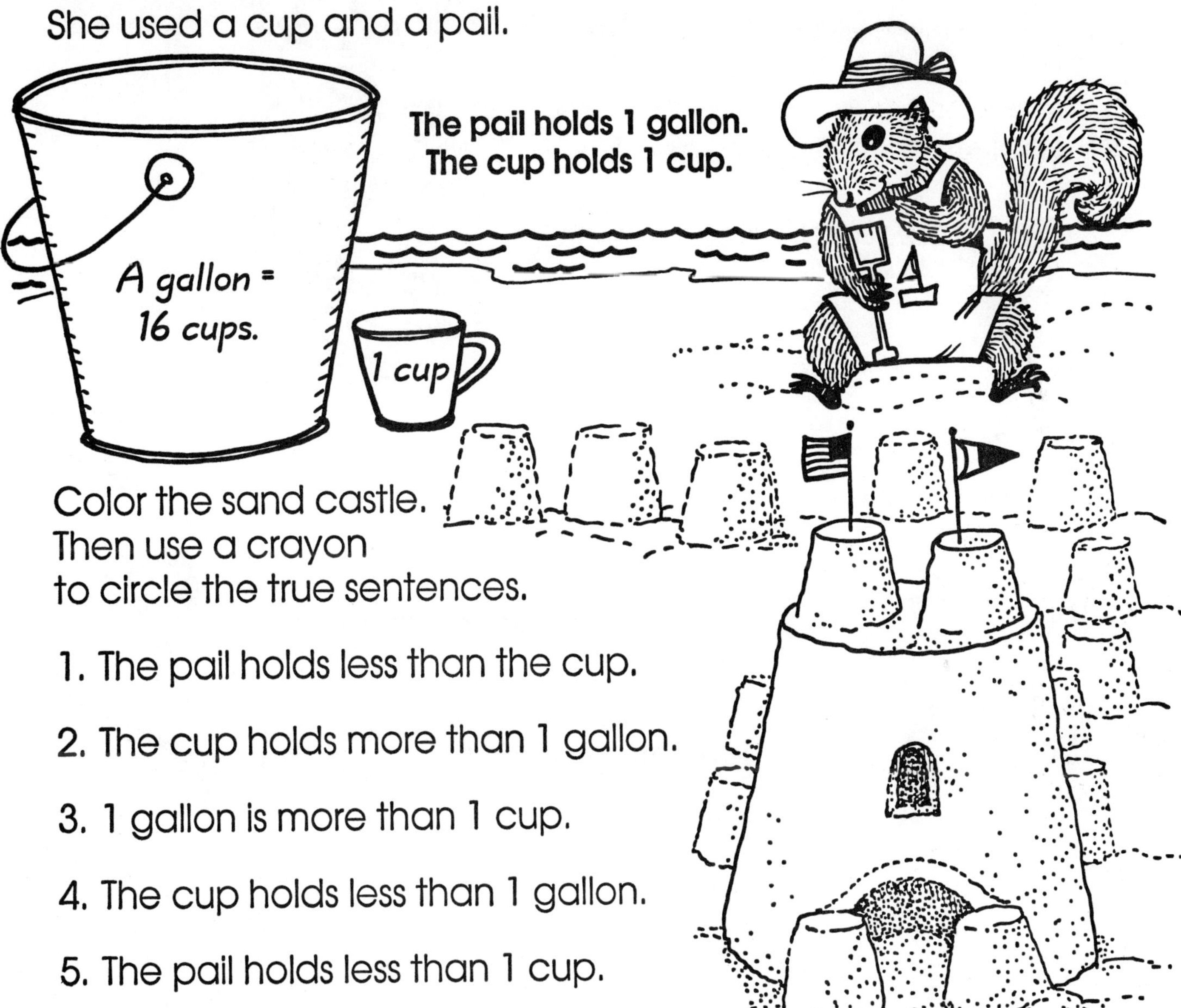

Color the sand castle. Then use a crayon to circle the true sentences.

1. The pail holds less than the cup.
2. The cup holds more than 1 gallon.
3. 1 gallon is more than 1 cup.
4. The cup holds less than 1 gallon.
5. The pail holds less than 1 cup.
6. 4 cups make a gallon.
7. 4 gallons equal 1 cup.
8. The big part of the sand castle was made with a cup.

Name ______________________________

The Long and Short of It

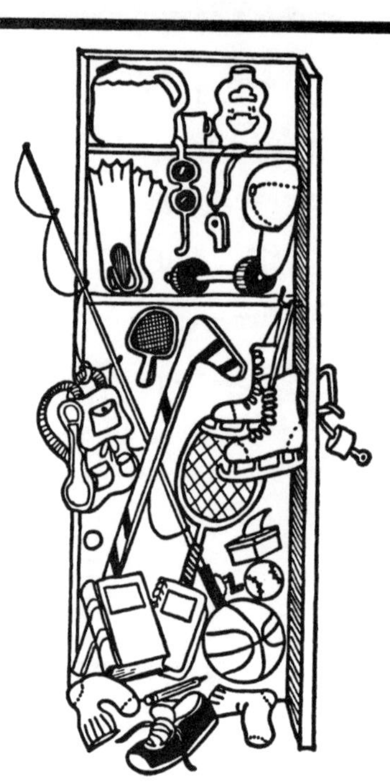

The athletes keep all their sports stuff in their lockers.

When they open their lockers, many things fall out!

Look at all the equipment!

Compare the sizes and weights.

Circle the best answers.

1. The  is **longer** **shorter** than the  .
2. The  is **longer** **shorter** than the  .
3. The  is **longer** **shorter** than the  .
4. The  is **longer** **shorter** than the  .
5. The  is **longer** **shorter** than the  .
6. The  is **longer** **shorter** than the  .

Name ______________________________

Use with page 55.

The Long and Short of It, continued

7. The  weighs **more** **less** than the 

.

8. The  weighs **more** **less** than the 

.

9. The  weighs **more** **less** than the  .

10. The  weighs **more** **less** than the  .

11. The  weighs **more** **less** than the 

.

12. The  weighs **more** **less** than the  .

13. The 

holds **more** **less** than the  .

14. The  holds **more** **less** than the 

.

15. The  holds **more** **less** than the  .

Color the pictures.

Name ______________________________

Use with page 54.

Math Words to Know

clock: A clock tells time. This clock shows a time of 7:30.

coins: Coins are worth an amount of money.

penny = 1¢

nickel = 5¢

dime = 10¢

quarter = 25¢

digit: Numbers have 1 or more digits. This number has 3 digits.

529

fraction: A fraction shows a part of something. This picture shows that $\frac{1}{3}$ of the pie is gone.

half: Two halves make a whole thing. This pizza is half gone. Half is a fraction: $\frac{1}{2}$

hundreds: The third place in a number is the hundreds place. In this number, the 8 is in the hundreds place.

853

length: Length measures something along a line.

The length of this worm is 3 inches.

measure: People measure to find out how long, big, or heavy something is or to find how much something holds. The lion uses a scale to measure his weight.

number line: A number line shows numbers in order along a line.

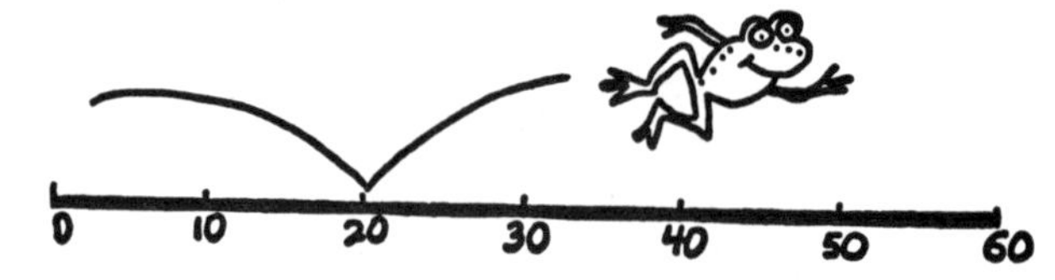

ones: The first place in a number is the ones place. In this number there are 6 ones.

276

place value: Place value tells the size of each digit in a number.

524

5 is in the hundreds place.
2 is in the tens place.
4 is in the ones place.

pound: The weight of things can be measured in pounds. A heavy pair of shoes weighs about 5 pounds.

quarter: There are 4 quarters in a whole thing. A quarter is a fraction, written $\frac{1}{4}$. One quarter of this pie is missing.

$\frac{1}{4}$

set: A set is a group of things you can count. This set has 6 things.

skip counting: To skip count, you skip over some numbers. Counting 2, 4, 6, 8, 10, 12 is skip counting by twos.

2 4 6 8 10 12

tens: The tens place is the second place in a number. In this number, 7 is in the tens place.

672

third: There are three thirds in a whole thing. One third is a fraction, written $\frac{1}{3}$. This shape is $\frac{1}{3}$ spotted.

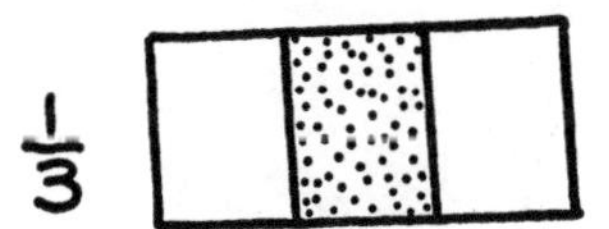

time: Clocks measure time. There are 60 seconds in a minute. There are 60 minutes in an hour. There are 24 hours in a day.

weight: Weight tells how heavy something is. Weight can be measured in ounces or pounds. This weight lifter is lifting 150 pounds.

Numbers & Counting Skills Test

Write a number to match the sets.

1. ________

2. ________

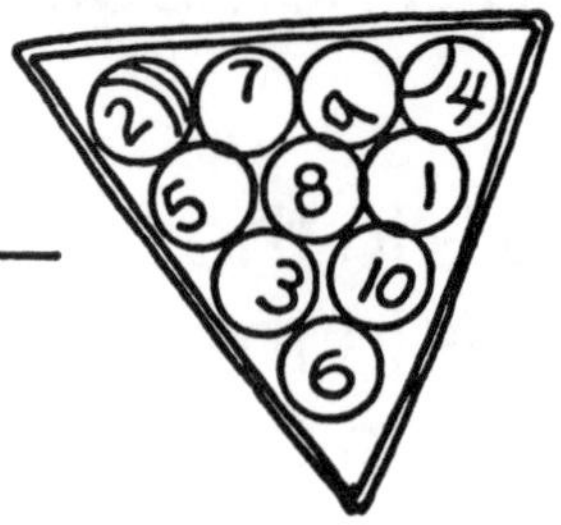

Put < or > in each box.

< means **less than**	> means **greater than**

3.  ☐

4.  ☐

5. Count by twos. Fill in the missing numbers.

2	4		8		12	14	

6. Count by fives. Fill in the missing numbers.

5		15			30		40

7. Count by tens. Fill in the missing numbers.

10			40		60			90	

Name ______________________________

Look at the picture below. Circle the right answer.

8. Who is second?	Todd	Joe	Deb
9. Who is fifth?	Ann	Sam	Todd
10. Who is first?	Deb	Bob	Sam
11. Where is Joe?	second	third	fourth
12. Where is Sam?	fourth	fifth	sixth

BOB DEB JOE ANN TODD SAM

Write the number for each of these.

13. four tens three ones ☐

14. six tens seven ones ☐

15. two tens no ones ☐

16. eight tens five ones ☐

Write the number for each word.

17. three ☐	18. zero ☐	19. nine ☐
20. seven ☐	21. twelve ☐	22. thirty ☐
23. eleven ☐	24. sixteen ☐	25. twenty ☐

Name ____________________________________

26. Circle the biggest number. 33 13 63 316

27. Circle the biggest number. 80 50 20 30 60

28. How much money? _____ ¢

29. How much money? _____ ¢

Put < or > in each box.

30. 32 ☐ 13

31. 11 ☐ 111

32. 41 ☐ 14

33. 69 ☐ 96

Write the missing number for each sentence.

34. ☐ = 70 + 7

35. ☐ = 90 + 2

36. ☐ = 40 + 8

37. ☐ = 100 + 10 + 1

38. ☐ = 70 + 9

39. Put a box around the clock that says 5:30.

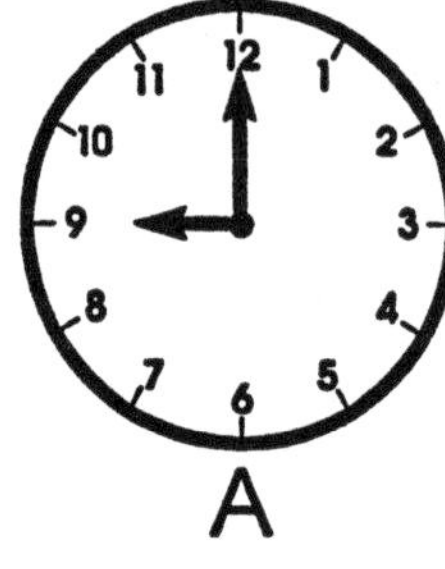

A

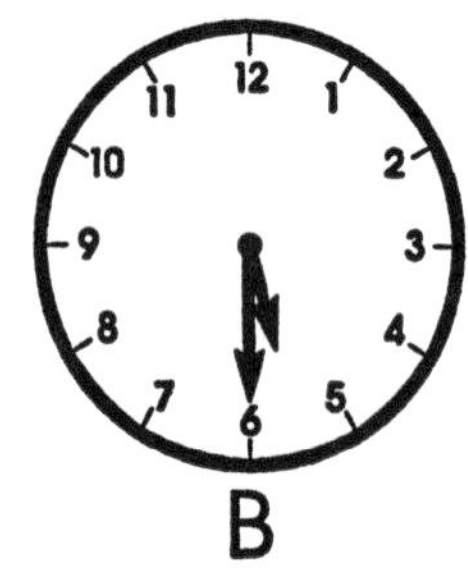

B

C

Write the time that each clock shows.

40.

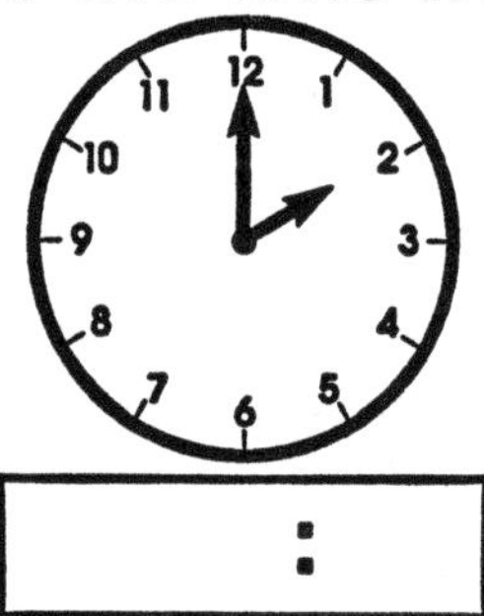

☐ :

41.

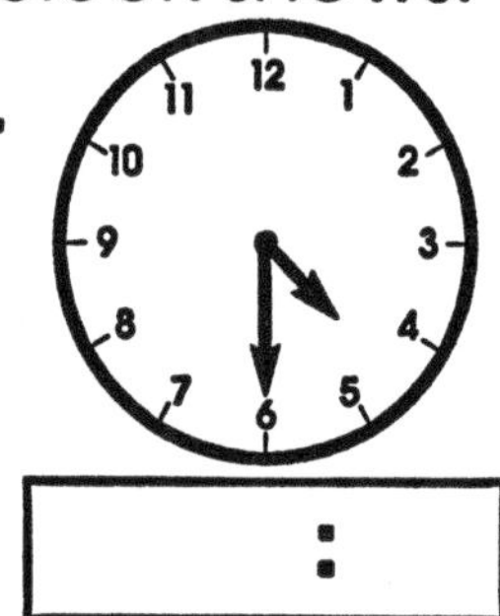

☐ :

42.

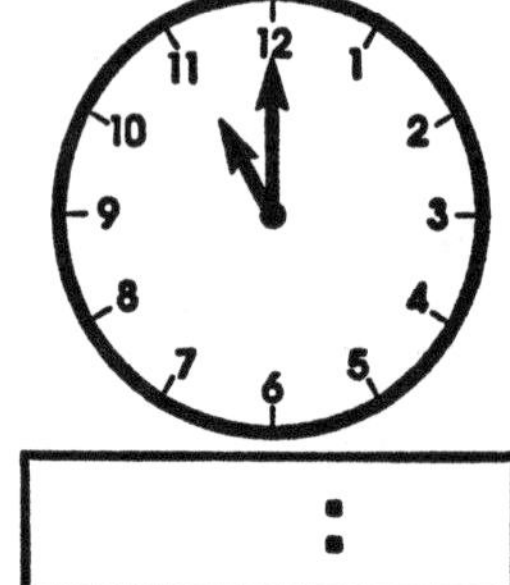

☐ :

Name ____________________

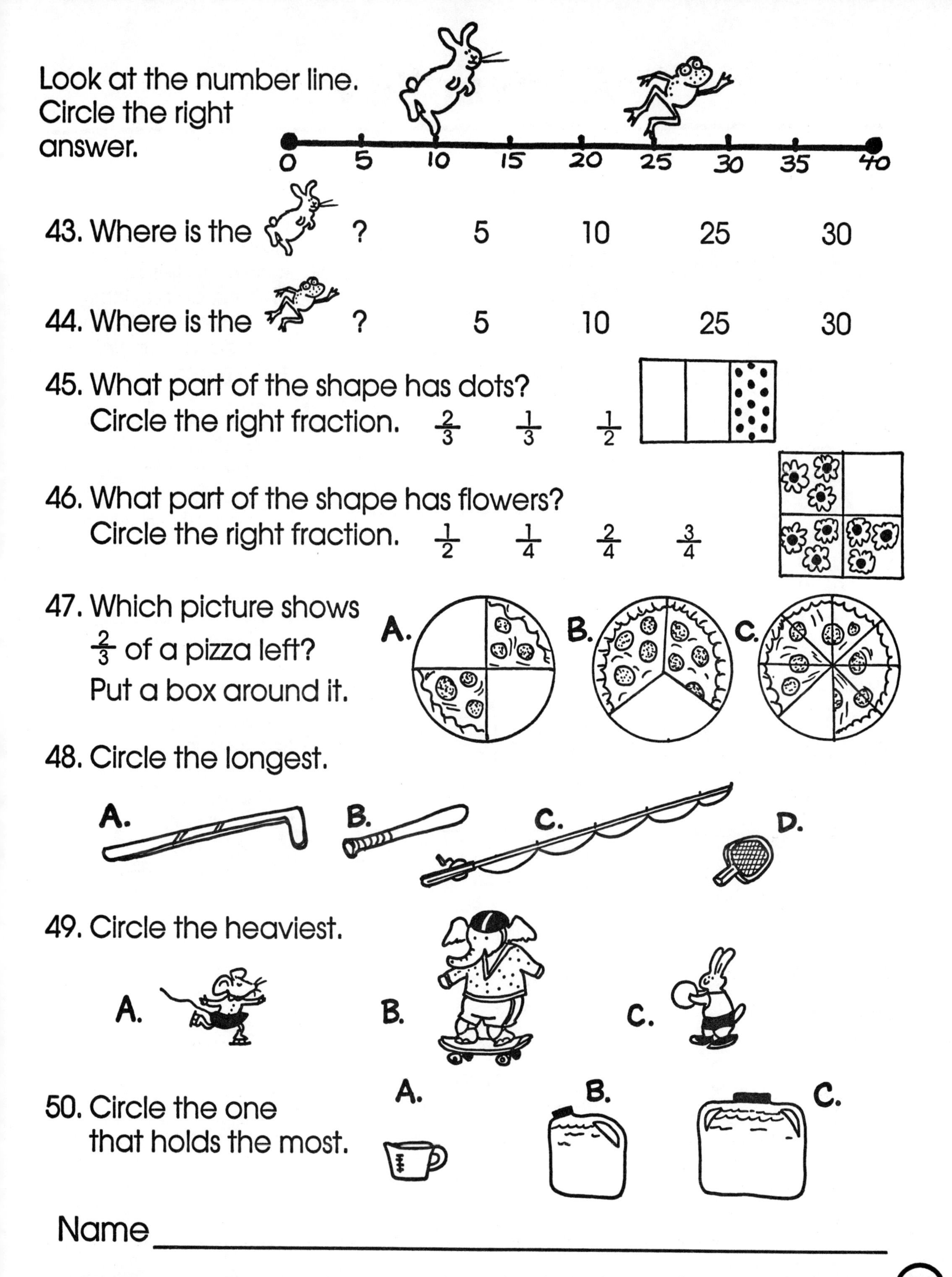

Look at the number line. Circle the right answer.

43. Where is the ? 5 10 25 30

44. Where is the ? 5 10 25 30

45. What part of the shape has dots?
Circle the right fraction. $\frac{2}{3}$ $\frac{1}{3}$ $\frac{1}{2}$

46. What part of the shape has flowers?
Circle the right fraction. $\frac{1}{2}$ $\frac{1}{4}$ $\frac{2}{4}$ $\frac{3}{4}$

47. Which picture shows $\frac{2}{3}$ of a pizza left?
Put a box around it.

A. B. C.

48. Circle the longest.

A. B. C. D.

49. Circle the heaviest.

A. B. C.

50. Circle the one that holds the most.

A. B. C.

Name ______________________________

Answer Key

Skills Test

1. 8
2. 10
3. <
4. >
5. Missing: 6, 10, 16
6. Missing: 10, 20, 25, 35
7. Missing 20, 30, 50, 70, 80, 100
8. Deb
9. Todd
10. Bob
11. third
12. sixth
13. 43
14. 67
15. 20
16. 85
17. 3
18. 0
19. 9
20. 7
21. 12
22. 30
23. 11
24. 16
25. 20
26. 316
27. 80
28. 70¢
29. 50¢
30. >
31. <
32. >
33. <
34. 77
35. 92
36. 48
37. 111
38. 79
39. B
40. 2:00
41. 4:30
42. 11:00
43. 10
44. 25
45. $\frac{1}{3}$
46. $\frac{3}{4}$
47. B
48. C
49. B
50. C

Skills Exercises

page 10

50 marbles

page 11

1. 3
2. 3
3. 5
4. 10
5. 4
6. 7

page 12

20 steps

page 13

See that students have filled in missing numbers correctly.

page 14

Draw lines from 2 to 56 to make jump ropes. Students count to 56.

page 15

Fill in the missing numbers to count 5, 10, 15, 20, 25, 30, 35, 40, 45, 50.
The highest number is 50.
The 2 players making high five are 25 and 30.

page 16

Missing numbers are 20, 30, 50, 70, 80, 100.

page 17

Look to see that students have connected the correct turtles to ribbons, colored the ribbons correctly, and colored the winning turtle green.

page 18

21 tiny fish at top—red
19 long-nosed fish on left—purple
7 large-mouthed fish in center—orange
11 fat fish to right—yellow

page 19

Students should add to the sets:
1 football
1 baseball cap
1 hockey stick
4 hockey pucks
3 golf balls
2 baseball bats

page 20

1. 40
2. 55
3. 75

Bottom—answers should all be "no."

page 21

1. 3
2. 9
3. 5
4. 2
5. 6
6. 1
7. 10

page 22

Look for:
goggles on frog,
yellow duck,
teeth on shark,
turtle's head,
hat on octopus,
and brown dog.

page 23

The team has more than 11 players.

1. 15	7. 90
2. 12	8. 62
3. 71	9. 17
4. 33	10. 8
5. 13	11. 50
6. 51	12. 26

page 24

Numbers should be placed on the fish in accordance with their size. The largest fish has 6000 on it and should be colored orange.

page 25

1. 80 + 9
2. 90 + 6
3. 100 + 30 + 3
4. 70 + 5
5. 60 + 9
6. 90 + 8

page 26

Across		Down	
A.	23	A.	20
C.	100	B.	30
D.	87	C.	17
E.	65	D.	85
F.	92	E.	66
G.	41	F.	91
H.	60	G.	40

page 27

Hidden clues are the bowling ball and pins. Sport is bowling.

page 28

1. 23	6. 30
2. 88	7. 88
3. 47	8. 16
4. 26	9. 88
5. 16	10. 30

page 29

red balls: 32, 123
blue balls: 14, 11
purple balls: 90, 111
brown ball: 66
orange ball: 29
green balls: 84, 50
yellow ball: 76
pink ball: 45

page 30

Left to right, student should draw on dancers:
1. green hoop
2. orange hoop
3. blue hoop
   purple hoop
   yellow hoop
4. red hoop

page 31

1. 5
2. 4
3. 3
4. 10
5. penguins
6. bears
7. snowboards

pages 32–33

a. 1	1. <
b. 24	2. >
c. 2	3. >
d. 18	4. >
e. 3	5. <
f. 11	6. =
g. 4	7. >
h. 3	8. <
i. 7	9. <
j. 6	10. >
k. 4	11. <
l. 5	
m. 3	

pages 34–35

Scores:
Gus 6
Gordy . . . 3
Gail 4
Gabby . . . 4
Gipper . . . 2

1. Gus
2. Gipper
3. Gus
4. Gipper

1. <
2. >
3. >
4. <
5. >
6. <
7. <
8. >
9. >
10. <

page 36

The monkey is surfing on a wave in the ocean.

page 37

Answers will vary.

page 38

See that students have drawn lines between the matching words and numbers.
There are 10 table tennis balls.

page 39

Answers will vary on estimates.
There are 10 mice.

pages 40–41

penny—1¢
nickel—5¢
dime—10¢
quarter—25¢

1. 10¢
2. 30¢
3. 90¢
4. $1.00

Total $2.30

page 42

1. pennies; 2¢
2. dimes; 60¢
3. quarter; 25¢
4. quarters; $1.00
5. dime; penny; 11¢
6. nickels; 10¢

Ben found the most money.

page 43

1. less
2. less
3. less
4. more
5. less
6. more

page 44

Check to see that students have drawn toppings in correct fractional sizes.

page 45

The paths students draw may vary, but must connect the following things:

Mike and Mac— red line joining:

- "one quarter"
- pictured $\frac{1}{4}$ of square
- pictured $\frac{1}{4}$ of circle
- coin—quarter
- $\frac{1}{4}$ fraction written
- "one fourth"
- time—15 minutes

Millie and Mary— blue line joining:

- "one half"
- pictured $\frac{1}{2}$ of square
- pictured $\frac{1}{2}$ of circle
- coin—half dollar
- $\frac{1}{2}$ fraction written
- time—30 minutes

page 46

1. box around $\frac{1}{2}$
2. circle around $\frac{1}{8}$
3. Answers will vary.
4. Answers will vary.

page 47

See that student has numbered clock correctly.

1. 12
2. 1
3. 1:00
4. 1:30

page 48

1. 5:00
2. 8:00
3. 10:30
4. 12:30
5. 3:00
6. 8:30

page 49

1. clock should show 3:30
2. clock should show 5:00
3. clock should show 6:30
4. clock should show 7:30

page 50

See that students have accurately filled in all dates on the calendar.

1. 3, 10, 17, 24
2. 4
3. 7
4. no
5. Check drawing.

page 51

1. $\frac{1}{2}$ inch
2. 1 $\frac{1}{2}$ inches
3. 2 inches
4. 1 inch
5. 5 inches
6. 1 $\frac{1}{2}$ inches
7. 1 inch
8. 2 inches
9. 1 $\frac{1}{2}$ inches
10. 2 $\frac{1}{2}$ inches

page 52

1. ounces
2. pounds
3. pounds
4. ounces
5. ounces
6. pounds
7. ounces
8. ounces

page 53

True sentences are 3 and 4.

pages 54–55

1. longer
2. shorter
3. shorter
4. longer
5. shorter
6. shorter
7. more
8. less
9. less
10. more
11. more
12. more
13. less
14. more
15. more